Leadership in Energy and Environmental Design

LEED® NC Sample Exam

New Construction

Second Edition

Meghan Peot, MEd and
Brennan Schumacher, LEED AP

Professional Publications Inc. • Belmont, CA

How to Locate and Report Errata for This Book

At Professional Publications, we do our best to bring you error-free books. But when errors do occur, we want to make sure you can view corrections and report any potential errors you find, so the errors cause as little confusion as possible.

A current list of known errata and other updates for this book is available on the PPI website at **www.ppi2pass.com/errata**. We update the errata page as often as necessary, so check in regularly. You will also find instructions for submitting suspected errata. We are grateful to every reader who takes the time to help us improve the quality of our books by pointing out an error.

LEED NC SAMPLE EXAM: NEW CONSTRUCTION
Second Edition

Current printing of this edition: 2

Printing History

edition number	printing number	update
1	1	New book.
2	1	New edition. Updated to LEED-NC Rating System v2.2.
2	2	Minor corrections.

Printed in the United States of America

PPI
1250 Fifth Avenue, Belmont, CA 94002
(650) 593-9119
www.ppi2pass.com

ISBN-13: 978-1-59126-108-7
ISBN-10: 1-59126-108-2

Library of Congress Control Number: 2007925069

Table of Contents

Preface and Acknowledgments

Our desire to play a larger role in influencing the performance of buildings and their impact on the environment has led to our work at Be Green Consulting™ (BGC). In addition to offering sustainable design services that provide a healthy and productive environment for occupants, BGC also provides educational services in green building and LEED®* certification processes through workshops and presentations.

The growing importance of LEED accreditation was our catalyst to work with PPI on *LEED NC Sample Exam*. This book was created to help individuals interested in this emerging field become LEED Accredited Professionals by providing a realistic exam for use as a study tool. There is no better way to evaluate one's knowledge prior to the actual exam than by taking this sample exam. Though the sample exam problems are similar to those on the actual exam, all problems are original to us and based on situations and questions we have encountered in our design work and teaching.

We wish to acknowledge those who have helped us update this second edition from the LEED-NC Rating System version 2.1 to version 2.2, including director of editorial Sarah Hubbard, director of production Cathy Schrott, and typesetter Kate Hayes. Thanks also to those others who assisted in the creation of the sample exam, including Nick Stecky, LEED AP, who provided insightful comments on our draft manuscript, and the PPI staff who saw the editorial and production work of the first edition through to the end: project editors Sean Sullivan and Scott Marley, proofreader Jenny Lindeburg, and typesetter and cover designer Amy Schwertman.

Meghan Peot, MEd
Brennan Schumacher, LEED AP

*The U.S. Green Building Council (USGBC) did not participate in the development and/or publication of this sample exam. The USGBC is not affiliated with PPI. LEED and USGBC are registered trademarks of the U.S. Green Building Council.

Introduction

About This Book

LEED NC Sample Exam: New Construction prepares the reader for the Leadership in Energy and Environmental Design (LEED) Professional Accreditation exam given by the U.S. Green Building Council (USGBC) in coordination with the Green Building Certification Institute (GBCI). The sample exam, based on the LEED-NC Rating System version 2.2, was designed to mirror the actual exam in length, content, and format. Each of the questions in this book has a detailed solution that explains the correct answer and provides insight into the LEED-NC Rating System and Reference Guide, which the exam is based on.

About the LEED Professional Accreditation Exam

The LEED Rating System is a voluntary, consensus-based national standard for developing high-performance, sustainable buildings. Members of the USGBC, representing all segments of the building industry, developed LEED and continue to contribute to its evolution.

Buildings become LEED certified, but individuals become LEED accredited. As defined by the USGBC in their publication *Candidate Handbook: LEED Professional Accreditation Exam*, the LEED-NC exam is designed to ensure that a successful candidate has the knowledge and skills necessary to participate in the design process, to support and encourage integrated design, and to streamline a building project's LEED application and certification process. Exam questions are designed to test an examinee's understanding of green building practices and principles and familiarity with LEED requirements, resources, and processes. Exam questions are based on the LEED-NC Rating System and Reference Guide, which are used in the certification process for new construction.

The exam tests for minimum competency in four subject areas, or sections, of the LEED certification process. These are identified by USGBC as follows.

- Section 1: Knowledge of LEED Credit Intents and Requirements
- Section 2: Coordinate Project and Team

- Section 3: Implement LEED Process
- Section 4: Verify, Participate in, and Perform Technical Analyses Required for LEED Credits

Each subject area is weighted to reflect its relative importance to the practice of a LEED Accredited Professional. These subject areas and their weighting are based on studies known as "task analyses" or "job analyses," which identify the knowledge, skills, and abilities needed of LEED Accredited Professionals working on LEED projects.

The exam is comprised of 80 randomly delivered multiple-choice questions that must be completed in 2 hours. Each question has four or more options to choose from. In some cases more than one option is correct, and the examinee must select all correct options to correctly answer the question. To ensure that an examinee's chances of passing remain constant regardless of the particular administration of the exam taken, the USGBC converts the raw exam score to a scaled score, with the total number of points set at 200 and a minimum passing score of 170. In other words, an examinee is not penalized if the exam taken is more difficult than usual. Instead, fewer questions must be answered correctly to achieve a passing score.

The LEED Professional Accreditation Exam currently offers three exam tracks: LEED for New Construction (LEED-NC), LEED for Commercial Interiors (LEED-CI), and LEED for Existing Buildings (LEED-EB). Each exam track is based on a specific version of the respective LEED Rating System. A candidate for LEED Professional Accreditation must choose one of these tracks to be tested under. Any one of these tracks will provide a successful candidate with the LEED AP credential. This LEED AP credential is applicable to all LEED Rating Systems.

Taking the Exam

The exam is administered by computer at any Prometric test site. Thomas Prometric is a third party testing agency with over 250 testing locations in the United States. When scheduling an exam, you must first register at www.gbci.org. Then you must go to the Prometric website at www.prometric.com/usgbc to schedule and pay for the exam. All cancellations and rescheduling must be done directly with Prometric. GBCI recommends that you show up a half hour before your exam appointment to check in and get settled.

A 15-minute tutorial is available for you to take before beginning the actual exam. Questions and answer choices are shown on a computer screen, and the computer keeps track of which answers you choose. You can skip questions, mark questions for later review, and change your answers until the time limit is up.

While taking the exam, it's better never to leave a question unanswered, even if it's one you want to skip and come back to. Instead, make your best quick guess right away, and mark the question to look at again later. If you decide on a different answer later, you can change it, but if you run out of time before getting to all your skipped questions, you will still have made a guess on each one.

Upon completing the exam, you will immediately receive your score. There is a 15-minute exit survey that completes the 2 hour and 30 minute exam experience. When exiting the examination room, an exam attendant will distribute your LEED Professional Accreditation Exam Score Report. If you pass, a LEED AP Certificate will be sent in the mail.

How to Use This Book

To get the most out of this book, treat it as a real exam. Don't look at the questions or solutions ahead of time. When you think you're ready to take the test, get a pencil and paper to record your answers with, sit down, set a timer for 2 hours, and solve as many questions as you can within the time limit.

After you've taken the sample exam, check your answers against the answer key. It contains a full explanation of each answer. Read the solutions in the back of the book, paying special attention to any questions you got wrong.

To do well on the real exam, you'll need two things: knowledge of the LEED-NC Rating System, and the ability to bring up that knowledge under time pressure. The *LEED NC Sample Exam* will help you improve both.

Sample Exam

1. A 55,000 sq ft, five-story office building is planned for the site of a large existing parking lot. The design includes bioswales to help treat stormwater and reduce runoff. Which of the following standards applies to SS Prerequisite 1, Construction Activity Pollution Prevention?

(A) USDA in the U.S. Code of Federal Regulations, Title 7, Volume 6
(B) U.S. EPA Document 832/R-92-005
(C) ASTM E1903-97 Phase II Environmental Site Assessment
(D) 2003 EPA Construction General Permit

2. A 30,000 sq ft office building in an eco-industrial park is being built to comply with SS Credit 8, Light Pollution Reduction. Which of the following pieces of information will be needed to provide a suitable exterior lighting design? (Choose four.)

(A) lamp lumens for exterior luminaires
(B) exterior pavement surfaces
(C) watts per square foot of exterior illumination
(D) location of property line
(E) interior lighting calculations indicating that the maximum candela value shall intersect opaque building interior surfaces
(F) watts per square foot of interior illumination

3. A firm is designing a Midwestern law office building with 15,000 sq ft of perimeter private offices, 30,000 sq ft of non-perimeter open offices, and 10,000 sq ft of non-regularly occupied spaces. Occupant comfort is a primary goal of the design criteria. To meet this goal, the design must meet criteria relating to ASHRAE 55, which refers to thermal comfort. Complying with ASHRAE 55 standards, however, may interfere with one of the following LEED-NC credits. Which one?

(A) EQ Credit 6.1, Controllability of Systems: Lighting
(B) EQ Credit 4.1, Low-Emitting Materials: Adhesives and Sealants

(C) EA Credit 5, Measurement and Verification
(D) EA Credit 1, Optimize Energy Performance

4. On projects over 50,000 sq ft, to comply with EA Prerequisite 1, Fundamental Commissioning of the Building energy systems, which one of the following can the commissioning agent be?

(A) the construction manager
(B) the mechanical engineer
(C) the mechanical engineer's subcontractor responsible for energy modeling
(D) an individual on the owner's staff

5. A 15,000 sq ft LEED-certifiable day care facility is being built. Indoor air quality, natural ventilation, and daylighting are primary design objectives. Specifications include low-VOC adhesives and sealants. Which of the following standards must be met to comply with EQ Credit 4.1, Low Emitting Materials: Adhesives and Sealants?

(A) ASHRAE 62.1-2004
(B) ASHRAE 62.2-2004
(C) SMACNA
(D) SCAQMD Rule #1168
(E) Green Seal's GS-11

6. One prerequisite in the LEED-NC Rating System involves meeting ASHRAE 62.1-2004. What does this standard refer to?

(A) heating requirements
(B) lighting levels
(C) ventilation requirements
(D) greenhouse gas regulations

7. A building owner is commissioning the renovation of a three-story, 78,000 sq ft facility, including a major renovation of the existing mechanical system. According to EA Credit 3, Enhanced Commissioning, there are certain tasks that must be performed by the commissioning agent. Commissioning agents must complete the following tasks. (Choose two.)

(A) Develop a systems manual to illustrate optimal operation of commissioned systems.
(B) Conduct one design review of the owner's design requirements, basis of design, and design documents prior to the mid-construction document phase.
(C) Review building operation 10 months after substantial completion.
(D) Verify that the requirements for training personnel and building occupants have been completed.

8. A 20,000 sq ft environmental learning center will be constructed on a 130,500 sq ft site with an existing asphalt parking lot. The center will be constructed from insulating concrete form (ICF) wall systems, and renewable resources such as small-diameter trees, bamboo, and cork will be used. Interpretive signage will teach visitors about the green building technologies and strategies used on-site. Which of the following would be appropriate credits to strive for within the LEED-NC Rating System? (Choose four.)

(A) EA Credit 2, On-Site Renewable Energy
(B) SS Credit 5.2, Site Development: Maximize Open Space
(C) SS Credit 4.1, Alternative Transportation: Public Transportation Access
(D) MR Credit 6, Rapidly Renewable Materials
(E) EA Credit 1, Optimize Energy Performance
(F) ID Credit 1, Innovation in Design

9. A project is planned in the northwest United States near Puget Sound, where the effect of runoff on the local salmon habitat is a major concern. The owners would like to promote biorention on-site and are considering a number of different low impact development strategies, including bioswales, porous pavement, and other approaches that mimic natural hydrologic conditions. Bioswales are used for

(A) water-efficiency in the landscape
(B) reducing the heat island
(C) erosion and sedimentation control
(D) stormwater management

10. Designers of a production facility are trying to earn an Innovation in Design credit in the LEED-NC Rating System by complying with building industry acoustical standards. Which of the following organizations provides such standards?

(A) American Society of Heating, Refrigeration and Air-Conditioning Engineers
(B) Architectural National Standards Institute
(C) Environmental Protection Agency
(D) American National Standards Institute

11. A construction project in Las Vegas has a primary purpose of educating the public on water conservation. The reduction in water usage will be 40% from baseline Energy Policy Act of 1992 standards. Which credits in the LEED-NC Rating System does the given project information provide points for?

(A) No points would be awarded because the building does not meet the criteria of WE Credit 1.1, Water Efficient Landscaping
(B) One point would be awarded for meeting the criteria of WE Credit 1.1, Water Efficient Landscaping
(C) Two points would be awarded for meeting the criteria of WE Credits 3.1 and 3.2, Water Use Reduction, and two points for meeting the criteria of ID Credits 1.1 and 1.2, Exemplary Performance and Education Demonstration
(D) Two points would be awarded for meeting the criteria of WE Credits 3.1 and 3.2, Water Use Reduction

12. Designers of a 30,000 sq ft school in a rural setting seek to earn SS Credit 6.2, Stormwater Design: Quality Control, in the LEED-NC Rating System. Incorporating which of the following green building elements in the design would help them attain this credit? (Choose three.)

(A) infiltration basin
(B) solar hot water system
(C) vegetated roof
(D) high-albedo concrete
(E) constructed wetland

13. Designers of a registered LEED project want to earn credit for providing resources for building occupants who bicycle to work. Although the project does not include showering and locker facilities, the business does offer showering facilities in an existing building within 100 ft of the new building's bicycle storage units. The project's scope includes a tenant space that is to be leased in the future. The design team is trying to confirm the FTE's for the project and is uncertain how to determine the FTE quantities of the future tenant spaces. Which of the following resources is the most definitive?

(A) USGBC Credit Interpretation Rulings
(B) local traffic study results
(C) *Environmental Building News*
(D) USGBC case studies of LEED-certified buildings

14. Owners of a research lab are planning a 150,000 sq ft facility that will include a Living Machine™ to comply with WE Credit 2, Innovative Wastewater Technologies in the LEED-NC Rating System. Which one of the following is NOT needed to document this point?

(A) number of occupants
(B) graywater volumes collected, if applicable
(C) flow rate on sprinkler heads
(D) number of workdays

15. A warehouse is being constructed near a wetland. Underground petroleum storage tanks were discovered during the design phase. The owners would like to know if their site is a brownfield and whether they can apply for grants to fund site cleanup. Which of the following CANNOT classify this site as a brownfield?

(A) local governmental agency
(B) civil engineer
(C) county court system
(D) EPA national or regional office

16. A restaurant's owners would like to reduce the use of potable water in the building. A mechanical engineer will submit documents to verify that potable water consumption has been reduced compared to baseline conditions. Which of the following must the LEED AP verify? (Choose three.)

(A) gallons per minute of urinal
(B) number of male occupants
(C) workdays per year
(D) square footage of building

17. A new headquarters building for a small bank is being constructed in the southwest United States. The design team is considering a roof with high emissivity. Emissivity is defined as the

(A) ratio of reflected solar energy to incoming solar energy
(B) ability of a material to shed infrared radiation or heat
(C) ratio of transmitted light to the total incident of light
(D) ratio of interior illuminance at a given plane to exterior light

18. The LEED-CI Rating System is most applicable to which of the following projects?

(A) a new 9,000 sq ft commercial real estate office building being constructed on a previously impacted site
(B) an existing commercial tenant space retrofitting the mechanical and electrical systems
(C) an existing tenant space previously occupied by a hair salon being renovated for a coffee shop
(D) an office complex being constructed for future commercial tenants

19. A city has just released a request for proposals. The request requires each design team to satisfy ID Credit 2, LEED Accredited Professional, in the LEED-NC Rating System. To meet this criterion,

(A) the design firm must employ one LEED AP
(B) the commissioning agent must be a LEED AP
(C) primary members of the design team must be LEED APs
(D) the LEED AP must be a principal member of the design team

20. The general contractor on a project suggests maximizing the amount of flyash used in making concrete while maintaining the structural integrity of the concrete as approved by the structural engineer. Which green building strategy will this contribute most directly toward?

(A) using recycled-content building materials
(B) using low-VOC building materials
(C) optimizing energy performance
(D) using salvageable building materials

21. It is important that the LEED consultant on each LEED project gather documentation for LEED-NC credits and inform all parties of their roles and responsibilities. The general contractor is in the best position to provide data and documentation for which of the following credits? (Choose three.)

(A) SS Credit 6.1, Stormwater Design: Quantity Control
(B) MR Credit 2, Construction Waste Management
(C) EA Credit 1, Optimize Energy Performance
(D) MR Prerequisite 1, Storage and Collection of Recyclables
(E) EQ Credit 4.1, Low-Emitting Materials: Adhesives and Sealants
(F) EQ Credit 3, Construction Indoor Air Quality (IAQ) Management Plan

22. The civil engineer on a LEED project is typically responsible for providing documentation for which of the following credits? (Choose two.)

(A) SS Prerequisite 1, Construction Activity Pollution Prevention
(B) SS Credit 4.3, Alternative Transportation: Low Emitting and Fuel Efficient Vehicles
(C) SS Credit 5.1, Site Development: Protect or Restore Habitat
(D) SS Credit 6.2, Stormwater Design: Quality Control
(E) SS Credit 7.2, Heat Island Effect: Roof
(F) SS Credit 8, Light Pollution Reduction

23. EA Credit 2, On-Site Renewable Energy, requires that at least 2.5% of the building's total energy use be supplied through on-site renewable energy systems. According to USGBC, which of the following is a renewable energy source?

(A) passive solar heating
(B) daylight harvesting
(C) photovoltaics
(D) ground source heat pumps

24. A project's baseline case is used to compare the outcomes of specific green building strategies in order to evaluate their impact. For which of the following credits is it necessary to use a baseline case?

(A) EA Credit 4, Enhanced Refrigerant Management, Option 1
(B) EQ Credit 6.1, Controllability of Systems
(C) EA Credit 6, Green Power
(D) EA Credit 1, Optimize Energy Performance, Option 1
(E) EQ Credit 5, Thermal Comfort

25. A local distribution company is designing a new warehouse. The design's primary focus is optimizing energy performance. To ensure that systems are in compliance with EA Credit 5, Measurement and Verification, which of the following must be continuously metered? (Choose four.)

(A) lighting systems and controls
(B) boiler efficiencies

(C) stormwater runoff volumes
(D) indoor water risers and outdoor irrigation systems
(E) daylight factor to lighting systems ratios
(F) building-related process energy systems and equipment

26. HVAC and refrigeration equipment can release refrigerants during operation and service. Some types of refrigerants are listed as ozone depleting compounds (ODCs). To reduce ODCs, production of what refrigerant was halted in 1995 in the United States?

(A) HCFC
(B) HFC
(C) CFC
(D) Halon

27. Responsibilities for LEED-NC documentation are commonly divided among the design and construction team during the design development phase. The mechanical engineer is typically responsible for providing documentation for which of the following credits? (Choose two.)

(A) EQ Prerequisite 1, Minimum Indoor Air Quality Performance
(B) EQ Credit 1, Outdoor Air Delivery Monitoring
(C) EQ Credit 4, Low-Emitting Materials
(D) EA Credit 2, On-Site Renewable Energy
(E) EA Credit 4, Enhanced Refrigerant Management

28. MR Credit 6, Rapidly Renewable Materials, requires that 2.5% of the total value of all building materials and products used in the project be rapidly renewable. According to USGBC, rapidly renewable plants are generally grown or raised and harvested within a

(A) 7-year cycle
(B) 10-year cycle
(C) 15-year cycle
(D) 25-year cycle

29. As a project's schematic design phase is being completed, the owner officially confirms that this will be a LEED project. A member of the design team will register the project with USGBC. LEED projects are registered with USGBC by

(A) setting up a meeting with a USGBC representative
(B) completing the registration online at www.usgbc.org
(C) completing USGBC's automated telephone registration
(D) printing and completing the online registration form and mailing it to USGBC

30. An environmental learning center is being constructed largely from local lumber. The owners would like to comply with MR Credit 7, Certified Wood, which requires that 50% of the virgin wood be FSC certified. To verify compliance with this credit, the LEED AP must know the costs associated with which of the following? (Choose two.)

(A) salvaged and refurbished wood-based materials
(B) rough carpentry
(C) the post-consumer recycled wood fiber portion of any product
(D) wood doors and frames

31. A 14,000 sq ft retail shop is opening in an existing building. The project will use black porous pavement systems made from pre-consumer recycled material. Which LEED-NC credits will the use of this building material help attain? (Choose two.)

(A) SS Credit 5, Site Development
(B) SS Credit 6, Stormwater Design
(C) SS Credit 7, Heat Island Effect
(D) SS Credit 8, Light Pollution Reduction
(E) MR Credit 4, Recycled Content
(F) MR Credit 5, Regional Materials

32. At what point in the LEED documentation process is it necessary to submit cut sheets containing VOC data in order to meet EQ Credit 4, Low-Emitting Materials?

(A) when requested by the commissioning agent
(B) when requested by OSHA
(C) when audited by USGBC
(D) during initial LEED certification submittal

33. Designers of a new building are researching ways to reduce greenhouse gases. The owners would like to locate the project close to public transportation, and want to find ways to provide alternative means of transportation. Which of the following statements INCORRECTLY defines a criterion for Low-Emission and Fuel Efficient Vehicles as specified in SS Credit 4.3, Alternative Transportation?

(A) Vehicles classified as zero emission by the California Air Resources Board.
(B) Vehicles that have achieved a minimum green score of 30 on the American Council for an Energy Efficient Economy vehicle rating guide.
(C) Vehicles powered by liquid natural gas.
(D) Vehicles powered by propane or compressed natural gas.
(E) Vehicles powered by gas-electric hybrid engines.

34. The LEED-NC Rating System uses the Energy Policy Act of 1992 to provide standards associated with

(A) site stormwater management
(B) plumbing fixtures
(C) ozone protection
(D) energy modeling

35. Plans for a high-performance 80,000 sq ft laboratory include the use of photoelectric daylight sensors. Which LEED-NC credit will the sensors contribute toward?

(A) EA Credit 1, Optimize Energy Performance
(B) EQ Credit 8, Daylight and Views
(C) SS Credit 8, Light Pollution Reduction
(D) EA Credit 2, On-Site Renewable Energy

36. USGBC has developed the *LEED-NC Reference Guide*. This resource is best described as

(A) a guide to all referenced standards with full abstracts regarding environmental policies
(B) a sustainable design guide and user's manual for the LEED-NC Rating System
(C) a sustainable material resource guide
(D) a green building directory of LEED Accredited Professionals

37. The design team wants to verify compliance with the requirements of SS Credit 2, Development Density and Community Connectivity. What project data should the LEED AP request of the design team to verify compliance? (Choose two.)

(A) building square footage
(B) property in acres
(C) number of full-time equivalent (FTE) building occupants
(D) development footprint

38. What are the benefits of LEED Certification? (Choose two.)

(A) on-line self certification of green building performance measures
(B) third party validation of a building's performance
(C) market exposure through the USGBC website
(D) discount attendance to all USGBC sponsored events

39. The USGBC website at www.usgbc.org has many resources for the green building industry. Which of the following can be found on the website? (Choose two.)

(A) Credit Interpretation Rulings
(B) green building materials
(C) lists of LEED Accredited Professionals arranged by area
(D) sample green building specifications

40. One standard referenced in the LEED-NC Rating System is ASHRAE 55-2004. Which building conditions does this standard refer to?

(A) temperature and lighting
(B) temperature and humidity
(C) humidity and lighting
(D) indoor air quality and lighting

41. MR Credit 4, Recycled Content, requires that the sum of post-consumer recycled content plus one-half of the pre-consumer content of materials with recycled content must constitute at least 10% of the total value of the building materials in the project. According to USGBC, post-consumer recycled content is defined as

(A) end-user waste that is reused
(B) end-user waste that has become feedstock for another product
(C) output from a process that has not been used as part of a consumer product
(D) end-user waste diverted from a landfill

42. A project's design team is using an integrated design approach to comply with green building standards set forth in the LEED-NC Rating System. Points are credited toward a project's LEED certification for

(A) having all design team members take the LEED-NC training workshop
(B) meeting minimum requirements of ASHRAE 62.1-2004 and approved Addenda using the ventilation rate procedure
(C) meeting ASHRAE 90.1-2004 or a more stringent local energy code
(D) not using refrigerants

43. Which of the following is NOT included in the requirements for ID Credit 1, Innovation in Design, as defined in the *LEED-NC Reference Guide*?

(A) Identify the intent of the proposed innovation credit.
(B) Identify the proposed requirements of compliance.
(C) Identify ongoing measurement and verification protocol.
(D) Identify submittals to demonstrate compliance.
(E) Identify design strategies that could be used to meet requirements.

44. The number of full-time equivalent (FTE) building occupants must be used consistently in calculations for which of the following LEED-NC credits? (Choose two.)

(A) SS Credit 4.2, Alternative Transportation
(B) SS Credit 4.4, Alternative Transportation
(C) WE Credit 3.1, Water Use Reduction
(D) EA Credit 1, Optimize Energy Performance

45. The commissioning agent for a LEED-registered project has broken out the scope of work relative to EA Prerequisite 1, Fundamental Commissioning of the Building Energy Systems, and EA Credit 3, Enhanced Commissioning. All the following are examples of enhanced commissioning tasks EXCEPT

(A) a commissioning authority independent of the design team shall conduct a review of the design prior to the construction documents phase
(B) an independent commissioning authority shall conduct a review of the construction documents near completion of the construction document development and prior to issuing the contract documents for construction

(C) an independent commissioning agent shall identify the commissioning team and its responsibilities

(D) an independent commissioning authority shall review the contractor submittals relative to the systems being commissioned

46. EA Prerequisite 2 refers to ASHRAE 90.1-2004, a common standard. Which of the following is ASHRAE 90.1-2004 concerned with?

(A) the standard method of testing building ventilation filters for removal efficiency

(B) building energy performance standards

(C) minimum levels of ventilation effectiveness and indoor air quality

(D) a standard for measuring air change effectiveness

47. The criteria for EQ Credit 3.1, Construction IAQ Management Plan: During Construction, require that equipment must comply with ASHRAE 52.2-1999. Which of the following does this common standard reference?

(A) a standard method of testing building ventilation filters for removal efficiency

(B) building energy performance standards

(C) minimum levels of ventilation effectiveness and indoor air quality

(D) a standard for measuring air change effectiveness

48. A project reusing an existing site is trying to earn an exemplary performance credit for SS Credit 5.1, Site Development: Protect or Restore Habitat. If the overall square footage of a previously developed site is 135,000 sq ft and the building footprint is 35,000 sq ft, which of the following must the project team complete in order to earn an ID Credit for exemplary performance?

(A) restore 35,000 sq ft of the existing site with native adaptive species

(B) double the amount of open space required by the credit

(C) restore 100,000 sq ft of the existing site with native adaptive species

(D) restore 75,000 sq ft of the existing site with native adaptive species

49. Windows play a vital role in the performance of a daylighting strategy in a building. One characteristic, visible transmittance, can be defined as the

(A) ratio of the total heat striking a surface to the heat transmitted

(B) ratio of the total light hitting a surface to the light transmitted

(C) ratio of the total energy hitting a surface to the light transmitted

(D) total light transmitted

50. Site development, grading, and clearing can cause significant erosion. Which of the following is an example of stabilization control for SS Prerequisite 1, Construction Activity Pollution Prevention, as stated in the *LEED-NC Reference Guide*?

(A) earth dikes

(B) silt fences

(C) sediment traps
(D) permanent seeding

51. Which of the following credits in the LEED-NC Rating System reference MERV Filter Ratings? (Choose two.)

(A) EQ Credit 1, Outdoor Air Delivery Monitoring
(B) EQ Credit 3.1, Construction IAQ Management Plan: During Construction
(C) EQ Credit 5, Indoor Chemical and Pollutant Source
(D) EA Credit 5, Measurement and Verification

52. Which two of the following are the responsibility of the owner during the design phase? (Choose two.)

(A) declaring that the building will be operated under a policy prohibiting smoking
(B) declaring and summarizing installation, operational design, and controls or zones for the CO_2 monitoring system
(C) declaring that the area dedicated to recycling will be easily accessible and accommodate the building's recycling needs
(D) declaring that parking will meet, but not exceed, minimum local zoning requirements

53. Which of the following is NOT a LEED-NC prerequisite?

(A) minimum energy performance
(B) CFC reduction in HVAC and refrigeration equipment
(C) site selection
(D) environmental tobacco smoke control

54. Who is responsible for assigning team roles to submit the LEED On-line templates?

(A) USGBC staff
(B) project architect
(C) owner
(D) project administrator

55. A project is nearing the end of its design phase, and the LEED consultant needs to verify that the owner will have a recycling plan in place and have a designated area for the collection and storage of recyclables. To comply with MR Prerequisite 1, which of the following recyclable materials must be collected on the site? (Choose three.)

(A) plastics
(B) corrugated cardboard
(C) paperboard
(D) metals

56. Which of the following is NOT an example of a rapidly renewable material?

(A) bamboo flooring
(B) wool carpet
(C) linoleum flooring
(D) FSC certified wood

57. The site selection criteria defined in the *LEED Reference Guide* for SS Credit 1, Site Selection, account for which of the following? (Choose three.)

(A) land documented as contaminated or classified as a brownfield by a local, state, or federal agency
(B) land specifically identified as habitat for any species on federal or state threatened or endangered lists
(C) previously undeveloped land less than 5 ft above the elevation of the 100-year flood as defined by the Federal Emergency Management Agency
(D) prime farmland as defined by the USDA in the U.S. Code of Federal Regulations

58. A LEED consultant is asked what the typical costs are for registering and certifying a building. The correct response is that these costs are based on

(A) project type
(B) building square footage
(C) location
(D) construction type

59. The intent of EQ Credit 6.2, Controllability of Systems, includes which of the following? (Choose two.)

(A) promote productivity, comfort, and well-being of occupants
(B) provide increasing levels of energy performance
(C) provide a high level of thermal ventilation
(D) control erosion to reduce negative impacts on water and air quality

60. What responsibilities of the civil engineer during the design submittal phase will support the LEED documentation process? (Choose two.)

(A) providing the spreadsheet calculation demonstrating that water-consuming fixtures reduce occupancy-based potable water consumption by 20% compared to baseline conditions
(B) providing a list of structural controls, including a description of the pollutant removal of each control and the percent annual rainfall treated
(C) declaring and demonstrating that stormwater management strategies result in at least a 25% decrease in the rate of stormwater runoff
(D) providing confirmation of the compliance path taken by the project (either NPDES compliance or local control standards)

61. Designers of a 48,000 sq ft, mixed-use project in New Mexico are trying to conserve water. Current design calls for a vegetated roof, composting toilets, and low-flow sinks and showers. The design team will also employ a drip irrigation system to reduce exterior site water usage. In order to verify compliance with WE Credit 1, Water Efficient Landscaping, what information is necessary to calculate the reduction in potable water used for site irrigation? (Choose three.)

(A) the microclimate factor
(B) the species factor
(C) a building plan demonstrating that vegetated roof areas constitute as least 50% of the total roof area
(D) calculations demonstrating that existing site imperviousness is less than or equal to 50%
(E) the density factor

62. A rural school plans to develop a greenfield site. The civil engineer would like to try to minimize the impact of development on the local ecosystem and has set up staging areas during construction. A point would be awarded within the LEED-NC Rating System for which one of the following credits?

(A) SS Credit 5.1, Site Development: Protect or Restore Habitat
(B) MR Credit 2, Construction Waste Management
(C) SS Prerequisite 1, Construction Activity Pollution Prevention
(D) EQ Credit 3.1, Construction IAQ Management Plan: During Construction

63. The design team has decided to pursue design strategies to meet the requirements of MR Credit 1, Building Reuse. The LEED AP must guide the design team with the development of technical analyses for which of the following? (Choose two.)

(A) structural elements in cubic feet
(B) structural elements in square feet
(C) shell elements in cubic feet
(D) shell elements in square feet
(E) exterior paving materials in cubic feet
(F) building footprint in square feet

64. A project design team is concerned about the amount of heat gain on the project site. The green building consultant has suggested that the team use materials that have high solar reflectance. Solar reflectance is

(A) the amount of light reflected by a material
(B) the ratio of reflected solar energy to incident solar energy
(C) emissivity
(D) the solar glazing on a window

65. The LEED On-line Credit Template is divided into which of the following four sections?

(A) Template Status, Manage Template, Required Documents, Documentation Status
(B) Required Documents, Documentation Status, Project Requirements, Credit Status
(C) Template Status, Manage Template, Credit Calculator, Project Details
(D) Manage Template, Required Documents, Team Role, Credit Status

66. USGBC offers many resources solely on the www.usgbc.org website. One such resource is the collection of LEED case studies. Which of the following kinds of information is NOT provided in these LEED case studies?

(A) complete LEED scorecard
(B) project photographs
(C) project statistics
(D) strategies and results

67. An architect meeting with an owner in a LEED project's programming and schematic phases has recommended that the project be registered as soon as possible. What are the benefits of early registration? (Choose two.)

(A) saving money on registration fees
(B) saving money on certification fees
(C) gaining access to the credit interpretation rulings database
(D) receiving two free Credit Interpretation Rulings
(E) maximizing the potential for achieving LEED certification

68. Which of the following materials is NOT considered to have pre-consumer recycled content?

(A) sawdust from a sawmill used to make composite board
(B) rubber from a tire plant used to make carpeting
(C) recycled pop bottles used to make carpeting
(D) spent grain from a brewery used as feed

69. The total materials cost figure may be derived from a nominal figure of 45% of total construction cost or a sum of actual material costs. The total materials cost figure must be used consistently in applying which of the following prerequisites or credits? (Choose two.)

(A) EQ Credit 4, Low-Emitting Materials
(B) MR Credit 3, Materials Reuse
(C) MR Credit 4, Recycled Content
(D) MR Prerequisite 1, Storage and Collection of Recyclables
(E) MR Credit 7, Certified Wood

70. A project design team is working to attain two points in WE Credit 1, Water Efficient Landscaping, and is looking for sources of graywater to use for site irrigation. Which of the following is a source of graywater?

(A) toilets
(B) food preparation sinks
(C) bathroom sinks
(D) waterless urinals

71. A design team is concerned about air quality during and after the construction of a hospital. SMACNA guidelines require that contractors cover ducts to prevent dust and other construction debris from entering the HVAC system during construction. Which of the following design team members will be impacted by the decision to follow SMACNA guidelines?

(A) electrical engineer and structural engineer
(B) commissioning agent and civil engineer
(C) general contractor and owner
(D) mechanical engineer and general contractor

72. A university in a warm climate is interested in reducing the energy consumption of the student union currently under construction. Which of the following strategies would help to accomplish this goal? (Choose two.)

(A) decreasing the lighting power density
(B) increasing the lighting power density
(C) decreasing the solar heat gain coefficient
(D) increasing the solar heat gain coefficient

73. A project team is considering purchasing 30% of a building's energy from a renewable energy resource as defined by the Center for Resource Solutions. Which credit or credits in the LEED-NC Rating System would this qualify for?

(A) EA Credit 2, On-Site Renewable Energy, and ID Credit 1, Innovation in Design
(B) EA Credit 6, Green Power
(C) EA Credit 1, Optimize Energy Performance
(D) no credit

74. The owners of a 19,000 sq ft community recreation center that is under construction want to measure the center's ongoing energy use. The project team is striving for EA Credit 5, Measurement and Verification. Due to the small size of the project, the team has decided to pursue Option B: Energy Conservation Measure Isolation of the International Performance Measurement and Verification Protocol (IPMVP). Which of the following should they use as the projected baseline energy use to comply with Option B of the IPMVP and earn EA Credit 5?

(A) calculated hypothetical baseline energy performance of the systems under measured post-construction operating conditions

(B) simulation of the baseline energy use under post-construction operating conditions

(C) energy use as determined by ASHRAE 90.1-2004 under post-construction operating conditions

(D) energy use as determined by the Department of Energy Buildings Consumption Survey (CBES)

75. The LEED AP is reviewing documents for compliance with EQ Credit 4.1, Low Emitting Materials: Adhesives and Sealants. Which of the following submittals and/or data sheets must the LEED AP review for compliance? (Choose two.)

(A) roofing adhesives

(B) fire stopping sealants

(C) stucco adhesives

(D) aerosol adhesives

76. Based on Energy Policy Act of 1992 standards and LEED default standards, what is the daily water use of ten FTE female occupants using a conventional water closet?

(A) three uses at 1.2 gallons

(B) thirty uses at 1.6 gallons

(C) forty uses at 1.2 gallons

(D) forty uses at 1.6 gallons

77. Which of the following statements about the Credit Interpretation Request (CIR) are true?

(A) Format the CIR as a letter including background information about the project and the questioned credit.

(B) Credit Interpretation Rulings guarantee a credit award.

(C) The CIR and ruling must be submitted with the LEED application.

(D) Include the credit name and your contact information when submitting a CIR.

78. The appeal is one possible step of the LEED certification and documentation process. Which of the following does NOT accurately reflect what should be included in an appeal?

(A) the LEED Project Checklist/Scorecard indicating projected prerequisites and credits and the total score for the project

(B) payment in the amount of $250 per credit or prerequisite appealed

(C) an overall project narrative including at least three project highlights

(D) LEED registration information such as project contact, type, size, number of occupants, and date of construction completion

79. Innovation in Design (ID) points for exemplary performance may be earned when the next incremental percentage threshold is achieved for which of the following credits? (Choose three.)

(A) SS Credit 5.2, Site Development: Maximize Open Space
(B) WE Credit 2, Innovative Wastewater Technologies
(C) EA Credit 1, Optimize Energy Performance
(D) EQ Credit 8.2, Daylight and Views: Views for 90% of Spaces
(E) EA Credit 6, Green Power

80. A project is using $100,000 worth of new wood and $50,000 worth of reclaimed wood flooring. What minimum value of wood must be used in order to earn credit for MR Credit 7, Certified Wood?

(A) $50,000 worth of new FSC certified wood
(B) $75,000 worth of new FSC certified wood
(C) $50,000 worth salvaged wood
(D) $75,000 worth of bamboo flooring

Answer Key

1. D
2. A, C, D, E
3. D
4. C
5. D
6. C
7. B, C
8. B, D, E, F
9. D
10. D
11. C
12. A, C, E
13. A
14. C
15. B
16. A, B, C
17. B
18. C
19. D
20. A
21. B, E, F
22. C, D
23. C
24. D
25. A, B, D, F
26. C
27. B, E
28. B
29. B
30. B, D
31. B, E
32. C
33. B
34. B
35. A
36. B
37. A, B
38. B, C
39. A, C
40. B
41. B
42. D
43. C
44. A, B
45. C
46. B
47 A
48. D
49. B
50. D
51. B, C
52. A, C
53. C
54. D
55. A, B, D
56. D
57. B, C, D
58. B
59. A, C
60. B, C
61. A, B, E
62. A
63. B, D
64. B
65. A
66. A
67. C, E
68. C
69. B, C
70. C
71. D
72. A, C
73. D
74. A
75. B, D
76. B
77. C
78. B
79. A, B, D
80. A

Solutions

1. ***The answer is*** **(D)** 2003 EPA Construction General Permit

2003 EPA Construction General Permit (CGP) "...outlines a set of provisions construction operators must follow to comply with the requirements of the National Pollution Discharge Elimination System (NPDES) stormwater regulations. The CGP covers any site one acre and above, including smaller sites that are part of a larger common plan of development or sale, and replaces and updates previous EPA permits." (www.epa.gov) A link to this document can be found at **www.ppi2pass.com/LEEDresources**. The CGP is the referenced standard in SS Prerequisite 1, Construction Activity Pollution Prevention, in the LEED-NC Rating System. While the CGP applies to sites based on size, all projects seeking LEED Certification must comply with the requirements of the CGP, or a more stringent local standard. Potential technologies for this prerequisite include both structural and stabilization control methods to minimize the negative impacts that erosion can have on water and air quality.

2. ***The answer is*** **(A)** lamp lumens for exterior luminaires
(C) watts per square foot of exterior illumination
(D) location of property line
(E) interior lighting calculations indicating that the maximum candela value shall intersect opaque building interior surfaces

The lighting designer will analyze the site conditions and the performance of the interior and exterior lighting systems to comply with the requirements of SS Credit 8. While all of the possible answers are important, not all are required data for demonstrating compliance with SS Credit 8. Answers (A), (D), and (E) all directly contribute to light pollution and (E) is the measurement of exterior lighting power densities. Exterior pavement surfaces have the potential to reflect light, however SS Credit 8 does not address reflected light off of pavement surfaces. The watts per square foot of interior illumination are important data for lighting power densities, however the watts per square foot of interior lighting systems do not directly impact light pollution. ASHRAE/IESNA 90.1-2004, Exterior Lighting

Section and RP-33-99 are referenced in the SS Credit 8, Light Pollution Reduction. SS Credit 8 provides requirements to reduce night sky pollution, light trespass across property lines, and excessive interior light transmittance through exterior windows.

3. ***The answer is*** **(D)** EA Credit 1, Optimize Energy Performance

ASHRAE 55 requires that occupied spaces that are mechanically heated and cooled must meet specific temperature and humidity conditions. Conditioning the air provides comfort, but it also increases energy consumption. The examinee must be aware of synergies and tradeoffs in applying green building strategies to the LEED-NC Rating System. This question represents a tradeoff where occupant comfort is given higher priority than optimizing energy performance.

4. ***The answer is*** **(C)** the mechanical engineer's subcontractor responsible for energy modeling

For projects over 50,000 sq ft, EA Prerequisite 1, Fundamental Commissioning of the Building Energy Systems, requires that the commissioning agent be independent of the project's design and construction teams. For projects smaller than 50,000 sq ft, the commissioning agent may be a qualified staff member of the owner, an owner's consultant to the project, or a qualified individual of the design or construction team that may have additional project responsibilities beyond commissioning. USGBC encourages that commissioning agents be independent third parties hired directly by the owner.

5. ***The answer is*** **(D)** SCAQMD Rule #1168

The South Coast Air Quality Management District is a government agency that works to create better air quality. The mission statement of SCAQMD is: "The South Coast AQMD believes all residents have a right to live and work in an environment of clean air and is committed to undertaking all necessary steps to protect public health from air pollution, with sensitivity to the impacts of its actions on the community and businesses." Additional information on the group can be found at www.aqmd.gov.

USGBC references SCAQMD Rule #1168 and Bay Area Air Quality Management District Regulation 8, Rule 51, which set VOC limits for sealants used as fillers, in Credit 4.1, Low-Emitting Materials Adhesives and Sealants, in the Indoor Environmental Quality category.

6. ***The answer is*** **(C)** ventilation requirements

The purpose of ASHRAE standard 62.1-2004 is to specify minimum ventilation rates and indoor air quality that will be acceptable to human occupants. The standard is intended to minimize the potential for adverse health effects. More details on this standard can be found at www.ashrae.org.

USGBC references this standard in EQ Prerequisite 1, Minimum Indoor Air Quality Performance.

7. ***The answer is*** **(B)** Conduct one design review of the owner's design requirements, basis of design, and design documents prior to the mid-construction document phase.
(C) Review building operation 10 months after substantial completion.

The Building Commissioning Association states that "The basic purpose of building commissioning is to provide documented confirmation that building systems function in compliance with criteria set forth in the project documents to satisfy the owner's operational needs." (www.bcxa.org)

USGBC makes two references to the commissioning process in the LEED-NC Rating System, EA Prerequisite 1, Fundamental Commissioning of the Building Energy Systems and EA Credit 3, Additional Commissioning.

According to EA Credit 3 requirements, answers (B) and (C) summarize two of the three tasks that must be completed by the commissioning agent. The third task the commissioning agent must complete is a review of the contractor submittals of commissioned systems for compliance with the owner's project requirements and basis of design. However, answers (A) and (D) are summaries of the commissioning tasks that can be completed by other members of the design team.

8. ***The answer is*** **(B)** SS Credit 5.2, Site Development: Maximize Open Space
(D) MR Credit 6, Rapidly Renewable Materials
(E) EA Credit 1, Optimize Energy Performance
(F) ID Credit 1, Innovation in Design

This question provides information that is relevant to the available choices, as well as some that isn't. The examinee will often need to read a question carefully and make an educated guess based on the clues in the question.

Bamboo, cork, and small-diameter trees are considered rapidly renewable resources. A high-efficiency wall system in ICF provides an opportunity to earn points in EA Credit 1, Optimize Energy Performance. Education and interpretive signage are relevant to ID Credit 1, Innovation in Design. A large site with a small building footprint allows the opportunity of SS Credit 5.2, Site Development: Maximize Open Space.

9. ***The answer is*** **(D)** stormwater management

Stormwater management is a focus of site development in green building. Low impact development strategies can be used to lessen the impact that buildings and developments have on local ecosystems.

USGBC references stormwater management specifically in two credits in the Sustainable Sites Category: SS Credits 6.1 and 6.2, Stormwater Design. These credits relate to the quantity of stormwater leaving the site and the ability to treat that water before it leaves the site.

10. ***The answer is*** **(D)** the American National Standards Institute

"The American National Standards Institute (ANSI) is a private, nonprofit organization (501(c)3) that administers and coordinates the U.S. voluntary standardization and conformity assessment system. The Institute's mission is to enhance both the global competitiveness of U.S. business and the U.S. quality of life by promoting and facilitating voluntary consensus standards and conformity assessment systems, and safeguarding their integrity." (www.ansi.org)

USGBC references many industry standards, and it is important to recognize common acronyms in the building industry.

11. ***The answer is*** **(C)** Two points would be awarded for meeting the criteria of WE Credits 3.1 and 3.2, Water Use Reduction, and two points for meeting the criteria of ID Credits 1.1 and 1.2, Exemplary Performance and Education Demonstration.

Baseline Energy Policy Act standards provide guidelines for interior water use, not landscaping water use, so options A and B are not valid. The project will earn two points for WE Credits 3.1 and 3.2 for having a 30% reduction in interior water use. In addition, the 40% water use reduction would qualify for an ID Credit 1, Innovation in Design for Exemplary Performance. This project would also qualify for ID Credit 1, Innovation in Design for Education, hence earning a second point for Innovation in Design. It is important to be familiar with some of the popular strategies for earning points in the Innovation in Design category.

The Energy Policy Act of 1992 is a reference standard that USGBC uses to establish criteria for WE Credit 3, Water Use Reduction. The act was established to conserve energy and water in buildings in the United States. It specifically addresses fixture flow requirements for water closets, urinals, showerheads, faucets, replacement aerators, and metering faucets.

12. ***The answer is*** **(A)** infiltration basin
(C) vegetated roof
(E) constructed wetlands

SS Credit 6.2, Stormwater Design: Quality Control, deals with the treatment of stormwater runoff. The three stormwater management techniques that answer this question help reduce the amounts of total suspended solids (TSS) and total phosphorous (TP) that leave a site. To qualify for this LEED-NC credit, it must be proven through design documents that the applied stormwater techniques are capable of removing 80% of TSS.

13. ***The answer is*** **(A)** USGBC Credit Interpretation Rulings

One resource of the LEED-NC Rating System through USGBC is Credit Interpretation Requests and Credit Interpretation Rulings. The rulings are the result of Credit Interpretation Requests that registered project design members have submitted to USGBC. These rulings serve as resources to clarify LEED-NC requirements and submittals. Each registered project is allowed two Credit Interpretation Requests and access to previous Credit Interpretation Rulings on the USGBC website. The previously accepted Innovations in Design can be viewed via the Credit Interpretation Rulings area of the USGBC website at www.usgbc.org. However, there is limited access to this resource for USGBC members, registered projects, and official USGBC workshop attendees.

14. ***The answer is*** **(C)** flow rate on sprinkler heads

The Water Efficiency category in the LEED-NC Rating System focuses on reducing the amount of potable water that projects use on the interior and exterior of buildings. The flow rate on sprinkler heads relates to exterior water consumption, so the data is not necessary for documentation of WE Credit 2, Innovative Wastewater Technologies. This credit includes strategies to reduce demand for potable water and generation of wastewater, and to treat the wastewater that is created on-site. The *LEED-NC Reference Guide* provides sample calculations to provide proper documentation of this credit, which includes variables that affect the quantity of wastewater created and saved. Although the LEED exam will not require the examinee to perform advanced calculations, it is helpful to be aware of the data required to perform these calculations.

15. ***The answer is*** **(B)** civil engineer

Brownfields are defined by the EPA as follows: "With certain legal exclusions and additions, the term 'brownfield site' means real property, the expansion, redevelopment, or reuse of which may be complicated by the presence or potential presence of a hazardous substance, pollutant, or contaminant." (www.epa.org)

USGBC provides incentive for the development of brownfields through SS Credit 3, Brownfield Redevelopment. These brownfields need to be defined by ASTM E1903-97 Phase II Environmental Site Assessment, or by a local, state, or federal agency.

16. ***The answer is*** **(A)** gallons per minute of urinal
(B) number of male occupants
(C) workdays per year

The calculation methodology in the *LEED-NC Reference Guide* for WE Credit 3, Water Use Reduction, reads as follows: "Create a spreadsheet listing the water-using fixture and frequency-of-use data. Frequency-of-use data includes the number of female and male daily uses, the duration of use, and the water volume per use."

USGBC defines potable water as water that is suitable for drinking and is supplied from wells or municipal water systems. Natural potable water around the globe is diminishing, but there are simple and cost-effective means of reducing how much potable water people use, including waterless urinals, dual flush toilets, metered faucets, composting toilets, and low-flow sinks, showers, and toilets.

17. ***The answer is*** **(B)** ability of a material to shed infrared heat

USGBC references emissivity in SS Credit 7.2, Heat Island Effect: Roof. Emissivity is used to determine the Solar Reflectance Index (SRI) of a roofing material. The USGBC uses the SRI as a measurement to determine whether or not roofing materials comply with SS Credit 7.2.

18. ***The answer is*** **(C)** an existing tenant space previously occupied by a hair salon renovated for a coffee shop

USGBC has developed several LEED rating systems as well as application guides to address various areas of the building industry. LEED-CI, for commercial interiors, allows building owners and businesses leasing spaces to certify their individual spaces with the LEED Rating System.

Other LEED rating systems include LEED-CS, for core and shell development, and LEED-EB, for existing buildings.

USGBC is constantly working to improve its applications of LEED as well as to develop other programs to address areas in the building industry that are in need of green building standards. It is important to be familiar with these systems. Check the USGBC website (www.usgbc.org) for up-to-date information on the status of these new rating systems.

19. ***The answer is*** **(D)** the LEED AP must be a principal member of the design team

USGBC provides one credit for having a LEED Accredited Professional as a principal member of the design team. This person must have successfully passed the LEED Professional Accreditation exam. Only one possible credit can be awarded, regardless of how many LEED Accredited Professionals are on the design team.

20. ***The answer is*** **(A)** using recycled-content building materials

Flyash is a waste material from coal-fired power plants and can partially replace cement in concrete applications. If not used in concrete, flyash would normally be landfilled. Varying percentages of flyash can be used in the concrete mix, depending on the building application. The structural engineer will specify the concrete mix based on the structural needs of the building application.

21. ***The answer is*** **(B)** MR Credit 2, Construction Waste Management
(E) EQ Credit 4.1, Low-Emitting Materials: Adhesives and Sealants
(F) EQ Credit 3, Construction Indoor Air Quality (IAQ) Management Plan

The general contractor plays an important role in the implementation and documentation of several LEED-NC credits. Others include MR Credit 3, Materials Reuse, MR Credit 4, Recycled Content, and MR Credit 5, Regional Materials.

It is important to know who is responsible for collecting and submitting information to provide documentation to USGBC along with the required submittal data.

22. ***The answer is*** **(C)** SS Credit 5.1, Site Development: Protect or Restore Habitat
(D) SS Credit 6.2, Stormwater Design: Quality Control

The civil engineer plays an important role in the documentation of several LEED-NC credits. Others include SS Credit 5.2, Site Development: Maximize Open Space, and SS Credit 6.l, Stormwater Design: Quantity Control.

It is important to know who is responsible for collecting and submitting information to provide documentation to USGBC along with the required submittal data. The fact that the question seeks only credits eliminates the possibility of (A), as SS Prerequisite 1, though documented by the civil engineer, is not actually a credit, but a prerequisite.

23. ***The answer is*** **(C)** photovoltaics

It is important to recognize the difference between renewable energy and energy efficiency. Several technologies are commonly mislabeled in this regard. A simple rule of thumb is that renewable energy, according to USGBC, actually creates power and/or thermal energy for use on-site, whereas energy efficiency saves or harvests energy. For example, a photovoltaic module actually creates electricity from the sun and thus is considered a form of renewable energy. A ground source heat pump is not an eligible source of renewable energy according to EA Credit 3. The *LEED-NC Reference Guide* states that earth-coupled HVAC applications that do not obtain significant quantities of deep-earth heat and use vapor-compression systems for heat transfer are not eligible for this credit.

24. ***The answer is*** **(D)** EA Credit 1, Optimize Energy Performance, Option 1

EA Credit 1, Optimize Energy Performance, Option 1, Whole Building Energy Simulation requires a baseline case to demonstrate a percentage improvement in the proposed building performance rating. Creating a baseline case of the project provides a point of reference for building owners to compare the effectiveness of green building strategies. In this case, ASHRAE 90.1-2004 establishes the baseline energy use.

25. ***The answer is*** **(A)** lighting systems and controls
(B) boiler efficiencies
(D) indoor water risers and outdoor irrigation systems
(F) building-related process energy systems and equipment

The energy performance requirements in EA Credit 5, Measurement and Verification, are based on the International Performance Measurement and Verification Protocol (IPMVP), Volume 1: Concepts and Options for Determining Energy and Water Savings (www.ipmvp.org). According to USGBC, "The IPMVP presents best-practice techniques for verifying savings produced by energy- and water-efficiency projects."

USGBC does not require continuous metering of stormwater runoff volumes or daylight factor to lighting systems ratios.

26. ***The answer is*** **(C)** CFC

CFC stands for chlorofluorocarbon. EA Prerequisite 3, CFC Reduction in HVAC&R Equipment, requires that all projects use CFC-free HVAC and refrigeration base-building and fire-suppression systems. LEED-NC has set parameters for buildings that contain old HVAC systems, but the federal government has banned the use of CFCs in most applications. The United States banned CFC production in 1995.

27. ***The answer is*** **(B)** EQ Credit 1, Outdoor Air Delivery Monitoring
(E) EA Credit 4, Enhanced Refrigerant Management

The mechanical engineer plays an important role in the documentation of several LEED-NC credits. Others include EQ Credit 2, Increased Ventilation Effectiveness, and EQ Credit 7, Thermal Comfort.

It is important to know who is responsible for collecting and submitting information to provide documentation to USGBC along with the required submittal data. The fact that the question is seeking only credits eliminates the possibility of (A) EQ Prerequisite 1, as this is documented by the civil engineer and is not actually a credit, but a prerequisite.

28. ***The answer is*** **(B)** 10-year cycle

The LEED-NC Rating System provides incentive for projects that are able to incorporate renewable resources in their construction materials and resources. This is accomplished in MR Credit 6, Rapidly Renewable Materials. USGBC and some industry standards use a 10-year cycle to define a rapidly renewable resource. It is important not to confuse this term with renewable energy.

29. ***The answer is*** **(B)** completing the registration online at www.usgbc.org

A portion of the exam is devoted to the LEED-NC implementation process. It is important to know about USGBC resources and processes as they are defined by the LEED-NC Rating System. In addition to its resources, the USGBC website is used for membership, personal accounts, and project registration.

30. ***The answer is*** **(B)** rough carpentry
(D) wood doors and frames

USGBC and many industry professionals see the Forest Stewardship Council label as a high standard in forestry harvesting practices. "The FSC is a nonprofit organization devoted to encouraging the responsible management of the world's forests." (www.fscus.org)

The calculation methodology in the *LEED-NC Reference Guide* for MR Credit 7, Certified Wood, reads: "Exclude salvaged and refurbished materials as well as the value of the post-consumer recycled wood fiber portion of any product. These exclusions ensure that applicants seeking the certified wood credit are not penalized for using non-virgin wood."

31. ***The answer is*** **(B)** SS Credit 6, Stormwater Design
(E) MR Credit 4, Recycled Content

The porous pavement will help attain SS Credit 6, Stormwater Design, by providing groundwater recharge and reducing stormwater runoff. It also has the ability to remove TSS. In addition, the porous pavement is made from pre-consumer recycled materials, which contributes toward MR Credit 4, Recycled Content. When one green building strategy applies to two or more credits, it is known as a synergy. This question also illustrates a tradeoff, when one green building strategy hinders another. The use of a black surface material on the site would limit the potential benefit that grid paving systems offer in the reduction of heat island effect.

32. ***The answer is*** **(C)** when audited by USGBC

USGBC uses the VOC content in materials as a designation for EQ Credit 4, Low-Emitting Materials. USGBC does not require the submittal of MSDS or cut sheets as part of the typical submittal process. However, USGBC does audit several of the credits as part of the review and certification process. It may be necessary to submit MSDS or cut sheets and additional information if audited by USGBC.

33. ***The answer is*** **(B)** Vehicles that have achieved a minimum green score of 30 on the American Council for an Energy Efficient Economy vehicle rating guide.

The requirements for SS Credit 4.3, Alternative Transportation: Low-Emission and Fuel-Efficient Vehicles define low emitting and fuel efficient vehicles as vehicles that have achieved a minimum green score of 40 on the American Council for an Energy Efficient Economy vehicle rating guide.

34. ***The answer is*** **(B)** plumbing fixtures

The 1992 Energy Policy Act addresses more than plumbing standards. It is meant to provide standards in energy and water conservation to reduce the dependency on foreign countries. USGBC uses this standard to establish a baseline for flow requirements for plumbing fixtures in WE Credit 3, Water Use Reduction.

35. ***The answer is*** **(A)** EA Credit 1, Optimize Energy Performance

Photoelectric daylight sensors help reduce energy consumption and increase the overall efficiency of the lighting system in a building. Take care not to confuse photoelectric sensors with photovoltaics.

Buildings can earn up to 10 points in EA Credit 1, Optimize Energy Performance.

36. ***The answer is*** **(B)** a sustainable design guide and user's manual for the LEED-NC Rating System

The *LEED-NC Reference Guide* is the definitive resource of the LEED-NC Rating System. It is a resource for LEED design and construction, providing sample calculations, diagrams, and design and environmental perspectives for each credit and prerequisite within the LEED-NC Rating System. It can be purchased through the USGBC website at www.usgbc.org.

37. ***The answer is*** **(A)** building square footage
(B) property in acres

The calculation methodology in the *LEED-NC Reference Guide* for SS Credit 2, Development Density and Community Connectivity, reads as follows: "Determine the total area of the project site and the total square footage of the building." These numbers allow the LEED AP to calculate the development density of a project.

38. ***The answer is*** **(B)** third party validation of a building's performance
(C) market exposure through the USGBC website

It is necessary to be knowledgeable about the LEED process from beginning to end. While technical skills are important in the application of LEED, in speaking with clients and the general public it is equally important to be well versed on the benefits of LEED Certification. Additional benefits of LEED Certification can be found on the USGBC website. (www.usgbc.org)

39. ***The answer is*** **(A)** Credit Interpretation Rulings
(C) lists of LEED Accredited Professionals arranged by area

USGBC uses its website as a primary means of communicating with the green building industry nationwide. Anyone can sign up for an account. It is important to be knowledgeable of the tools available on this website. Go to www.usgbc.org for more information.

40. ***The answer is*** **(B)** temperature and humidity

EQ Credit 7, Thermal Comfort, references ASHRAE 55-2004 to ensure that building occupants are provided a comfortable thermal environment that supports productivity and well being. "This standard specifies the combinations of indoor space environment and personal factors that will produce thermal environmental conditions acceptable to 80% or more of the occupants within a space. The environmental factors addressed are temperature radiation, humidity, and air speeds; the personal factors are those of activity and clothing." (www.ashrae.org)

41. ***The answer is*** **(B)** end-user waste that has become feedstock for another product

USGBC makes reference to recycled content in MR Credit 4, Recycled Content. There are two basic types of recycled content: post-consumer and pre-consumer. USGBC provides a higher weighted average for post-consumer recycled content, as these materials have reached the end user.

USGBC defines post-consumer waste as material generated by households or by commercial, industrial, and institutional facilities in their role as end-users of a product that can no longer be used for its intended purpose. An example of a material with post-consumer recycled content is carpet that is made from beverage bottles that were used in a home and then recycled. Pre-consumer waste is industrial waste that becomes feedstock for another industrial process. Industrial waste that becomes feedstock in the same process is not considered recycled content according to the LEED-NC Rating System.

42. ***The answer is*** **(D)** not using refrigerants

The examinee must know all prerequisites in the LEED-NC Rating System. Prerequisites do not themselves earn points in the rating system, but are simply requirements.

This question references the requirements for EA Credit 4, Enhanced Refrigerant Management.

43. ***The answer is*** **(C)** Identify ongoing measurement and verification protocol.

This questions tests knowledge of the process for ID Credit 1, Innovation in Design. The *LEED-NC Reference Guide* states the requirements of ID Credit 1, Innovation in Design, as: "In writing, identify the intent of the proposed innovation credit, the proposed requirements for compliance, the proposed submittals to demonstrate compliance, and the design approach (strategies) that might be used to meet the requirements." There is currently no requirement to identify ongoing measurement and verification protocol.

44. *The answer is* **(A)** SS Credit 4.2, Alternative Transportation
(B) SS Credit 4.4, Alternative Transportation

Full-time equivalent (FTE) building occupants must be used consistently in calculations in the LEED-NC Rating System. FTE is equal to the number of worker hours divided by eight. (Eight hours represents a full-time workday). According to the *LEED-NC Reference Guide*, a full-time worker has an FTE value of 1.0, and a half-time worker has a value of 0.5.

45. *The answer is* **(C)** an independent commissioning agent shall identify the commissioning team and its responsibilities

The LEED-NC Rating System and *LEED-NC Reference Guide* provide a clear description of the scope of work associated with commissioning services in EA Prerequisite 1, Fundamental Commissioning of the Building Energy Systems, and EA Credit 3, Enhanced Commissioning. Identification of the commissioning team and its responsibilities is not part of the scope of work.

46. *The answer is* **(B)** building energy performance standards

The LEED-NC Rating System references ASHRAE 90.1-2004, which provides minimum standards for energy-efficient design of buildings in EA Prerequisite 2, Minimum Energy Performance, and EA Credit 1, Optimize Energy Performance. Included in ASHRAE 90.1-2004 are regulated building loads such as building envelope, HVAC, service water heating, power, lighting, and other equipment, including all permanently wired electrical motors. ASHRAE 90.1-2004 does not set standards for low-rise residential buildings. For more information, go to www.ashrae.org.

47. *The answer is* **(A)** a standard method of testing ventilation filters for removal efficiency

USGBC references ASHRAE 55.2-1999 in EQ Credit 3, Construction Indoor Air Quality (IAQ) Management Plan. ASHRAE 55.2-1999 provides methods for measuring the performance of air cleaners based on the ability of a device to remove particles from the airstream and the device's resistance to airflow. For more information, go to www.ashrae.org.

48. *The answer is* **(D)** restore 75,000 sq ft of the existing site with native adaptive species

The two types of innovation strategies that qualify for ID Credit 1, Innovation in Design are Exemplary Performance and Innovative Performance. The *LEED-NC Reference Guide* defines Exemplary Performance strategies as those that greatly exceed the requirements of existing LEED credits. USGBC has set criteria for Exemplary Performance in SS Credit 5.1, Site Development: Protect or Restore Habitat by protecting or restoring 75% of the site area with native or adaptive vegetation on previous developed or graded sites. This will exceed the base criteria set at 50%.

In this problem, the overall square footage of the previously developed site is 135,000 sq ft and the building footprint is 35,000 sq ft, therefore according to the requirements for SS Credit 5.1, the total site area is 100,000 sq ft. Protecting or restoring 75% of the total site area means 75,000 sq ft must be restored.

49. ***The answer is*** **(B)** ratio of the total light hitting a surface to the light transmitted

Daylighting has proven to be a cost-effective means of providing an interior environment that promotes the well-being of building occupants. USGBC makes reference to daylight in EQ Credit 8.1, Daylight and Views: Daylight 75% of Spaces. This credit considers the value of visible light transmittance when calculating the glazing factor in a building. The *LEED-NC Reference Guide* defines the glazing factor as a ratio of the interior illuminance at a given point on a given plane to the exterior illuminance under overcast sky conditions. There are many things that influence this factor, such as window placement, window geometry, floor area, and total window area. The LEED-NC criterion for the glazing factor has been established at 2%. There are several ways to accomplish this calculation: through manual calculations, with daylight modeling with physical models, or with computer models.

50. ***The answer is*** **(D)** permanent seeding

This question requires the examinee to know the difference between stabilization control and structural control measures. Permanent seeding, temporary seeding, and mulching are all examples of stabilization control measures. Earth dikes, silt fences, sediment traps, and sediment basins are examples of structural control measures. If the examinee does not know the difference between stabilization and structural control measures, then a process of elimination could be employed. Each of the first three control methods is a structure or element that creates a structure, whereas the last option is a planting technique that stabilizes the soil.

51. ***The answer is*** **(B)** EQ Credit 3.1 Construction IAQ Management Plan: During Construction
(C) EQ Credit 5, Indoor Chemical and Pollutant Source

MERV is an acronym for minimum efficiency reporting value. It is based a filter's ability to remove particles and the filter's resistance to air flow. This rating comes from the ASHRAE Standard 52.2-1999, which provides a method for testing the performance of air filters.

52. ***The answer is*** **(A)** declaring that the building will be operated under a policy prohibiting smoking
(C) declaring that the area dedicated to recycling will be easily accessible and accommodate the building's recycling needs

This question requests that the examinee be knowledgeable about the roles of the owner during the design phase. It is important to know who is responsible for specific credits, along with knowing the required submittal data.

Choices (B) and (D) are the responsibilities of the mechanical engineer and the architect, respectively.

53. ***The answer is*** **(C)** site selection

SS Credit 1, Site Selection, is the only credit of these choices. The examinee must be able to tell which are LEED-NC credits and which are prerequisites. One way to gain the required knowledge of the LEED-NC Rating System is to memorize every LEED-NC credit and prerequisite.

54. ***The answer is*** **(D)** project administrator

The USGBC states that the project administrator assigns a team role to a project team member. This allows the member to submit templates and have on-line access to the project's assigned credits. More information on managing team roles can be found in the Team Admin area of the help section of http://leedonline.usgbc.org.

55. ***The answer is*** **(A)** plastics
(B) corrugated cardboard
(D) metals

USGBC requires all projects to have a dedicated on-site storage space for recycling at a minimum paper, corrugated cardboard, glass, plastics, and metals. This is one of the few LEED-NC credits and prerequisites that refer to the operation and maintenance of the building after construction.

56. ***The answer is*** **(D)** FSC certified wood

USGBC defines rapidly renewable materials as those that are made from plants and resources that have a harvest life cycle of 10 years or less. Following this definition, both bamboo and wool carpet are considered rapidly renewable materials. Some types of linoleum are also considered rapidly renewable materials, such as those made from linseed oil, rosins, wood flour, jute, and limestone. FSC certified wood is left as the best answer. Although LEED provides a point for using FSC certified wood in MR Credit 6, it typically has a harvest cycle longer than 10 years.

57. ***The answer is*** **(B)** land specifically identified as habitat for any species on federal or state threatened or endangered lists
(C) previously undeveloped land less than 5 ft above the elevation of the 100-year flood as defined by the Federal Emergency Management Agency
(D) prime farmland as defined by the USDA in the U.S. Code of Federal Regulations

Choice (A) pertains to SS Credit 3, Brownfield Redevelopment, which encourages projects to develop on and remediate the contamination of brownfield sites. All other choices pertain to SS Credit 1, Site Selection.

58. ***The answer is*** **(B)** building square footage

The only LEED-NC specific costs for projects come from the actual registration and certification costs, which are based on overall square footage of the building. Project registration costs are now fixed costs and certification costs are based on square footage. Also, certification costs can be broken into two submittals: Design Review and Construction Review. A more specific breakdown is as follows.

LEED-NC Registration

fixed rate of $450 for members, $650 for non-members

LEED-NC Certification Costs

review	area (sq ft)	member price	non-member price
design			
	0–50,000	$1250	$1500
	50,000–500,000	$0.025/sq ft	$0.030/sq ft
	500,000 and bigger	$12,500	$15,000
construction			
	0–50,000	$500	$5000
	50,000–500,000	$0.01/sq ft	$0.015/sq ft
	500,000 and bigger	$750	$7500
combined design and construction			
	0–50,000	$1750	$17,500
	50,000–500,000	$0.035/sq ft	$0.045/sq ft
	500,000 and bigger	$2250	$22,500

59. ***The answer is*** **(A)** promote productivity, comfort, and well-being of occupants
(C) provide a high level of thermal ventilation

Choices (A) and (C) are included in the intent of EQ Credit 6.2, Controllability of Systems. Choice (B) pertains to the intent of EA Credit 1, Optimize Energy Performance, and choice (D) pertains to SS Prerequisite 1, Erosion and Sedimentation Control.

60. ***The answer is*** **(B)** providing a list of structural controls, including a description of the pollutant removal of each control and the percent annual rainfall treated.
(C) declaring and demonstrating that stormwater management strategies result in at least a 25% decrease in the rate of stormwater runoff.

This question requires that the examinee be knowledgeable about the roles of the civil engineer for the LEED documentation process. It is important to know who is responsible for specific credits, the required submittal data, and during which phase each credit and prerequisite must be submitted.

61. ***The answer is*** **(A)** the microclimate factor
(B) the species factor
(E) the density factor

This question concerns the information needed to calculate the percentage of water efficiency based on the landscaping design. These factors are necessary in order to calculate the landscape coefficient, which will be needed by the civil engineer or landscape architect in calculating the water efficiency of the site. This question is typical in that there is information provided that is not relevant to the answer. For example, "vegetated roof, composting toilets, and low-flow sinks and showers" do not relate to the landscaping. On the actual test, it will be necessary to distinguish between information needed to answer the question and information meant to confuse.

62. ***The answer is*** **(A)** SS Credit 5.1, Site Development: Protect or Restore Habitat

USGBC addresses potential negative impacts that development can have on habitat through SS Credit 5.1, Site Development: Protect or Restore Habitat. This credit requires that projects developing on greenfield sites maintain development boundaries to protect open space. One strategy involves setting up staging areas. On areas that have been previously developed, LEED-NC requires that the project restore 50% of degraded habitat site areas, excluding the building footprint. It may appear that SS Prerequisite 1, Construction Activity Pollution Prevention, would be applicable; however this is a prerequisite and would not earn points in the LEED process.

63. ***The answer is*** **(B)** structural elements in square feet
(D) shell elements in square feet

Documentation of MR Credit 1, Building Reuse, requires calculations based on primary building elements of the existing building. Structural and shell elements are calculated in square feet. These data are required in order to calculate percentages of structural and shell elements reused in a building, which are needed to find the total percentage of building reuse.

Non-shell elements are also calculated in square feet in MR Credit 1.3, Building Reuse.

64. ***The answer is*** **(B)** the ratio of reflected solar energy to incident solar energy

Solar reflectance is defined as the ability of a surface material to reflect sunlight and is a ratio of the reflected solar energy to the incoming solar energy, expressed as a number between 0 and 1. Higher values of solar reflectance result in better control of heat gain. Solar reflectance includes the visible light, infrared heat, and ultraviolet wavelengths. Solar reflectance is also known as albedo.

Emissivity is commonly mistaken for solar reflectance. They have similar characteristics, but emissivity pertains to the ability of a surface to emit or shed infrared heat.

65. ***The answer is*** **(A)** template status, manage template, required documents, documentation status

The USGBC states the LEED On-line Credit Template section is the main area where the project administrator or team will document the necessary information for attempted credits.

More information on the Credit Template can be found at http://leedonline.usgbc.org.

66. ***The answer is*** **(A)** complete LEED scorecard

LEED scorecards can be found on the USGBC website under the Certified Project List. LEED case studies provide overviews of LEED-certified projects, they highlight green building strategies, and they may provide general cost data, square footages, project team information, and even pictures. LEED case studies provide insight on projects that have successfully implemented green building strategies based on factors such as region, construction type, building type, and project size.

67. ***The answer is*** **(C)** gaining access to the credit interpretation rulings database
(E) maximizing the potential for achieving LEED certification

LEED project registration provides project teams with access to software tools, LEED Online, and the credit interpretation rulings database. Access to this information early in a project (such as during the programming and schematic phases) allows project teams to better plan their LEED strategies, thus maximizing the potential for LEED certification.

The registration information is used by USGBC to respond to credit interpretation requests, track certification progress, and compile green building costs and trends.

68. ***The answer is*** **(C)** recycled pop bottles used to make carpeting

This question makes reference to recycled content as defined in MR Credit 4, Recycled Content. There are two basic types of recycled content: post-consumer and pre-consumer. USGBC provides a higher weighted average for post-consumer recycled content, as these materials have reached the end user.

Pre-consumer recycled content is industrial waste that becomes feedstock for another industrial process. Industrial waste that becomes feedstock in the same process is not considered recycled content according to the LEED-NC Rating System.

69. ***The answer is*** **(B)** MR Credit 3, Materials Reuse
(C) MR Credit 4, Recycled Content

The total materials cost figure may be derived from a nominal number of 45% of construction cost or from a sum of actual material costs. This total cost value is used in MR Credits 3, 4, 5, and 6. This figure represents the cost of building materials only. It does not include influences such as overhead, profit, rental fees, or labor. It also excludes electrical and mechanical systems. (These systems can be included as an Innovation in Design.) Any furnishings that are included must be included consistently throughout the material calculations.

70. ***The answer is*** **(C)** bathroom sinks

Graywater systems are becoming more common as water demands are becoming a higher priority on the national level. USGBC makes reference to such terms as potable water, graywater, and blackwater. Graywater can be reused for non-potable purposes and can be captured from rainwater and sinks and then filtered and applied to site irrigation or toilet flushing. Graywater sources include water from lavatory sinks, showers, bathtubs, and washing machines. All other possible answers, along with wastewater containing organic, toxic, or hazardous materials from sinks, are considered blackwater. Blackwater is also known as raw sewage and requires significant treatment before it can be reused.

71. ***The answer is*** **(D)** mechanical engineer and general contractor

USGBC provides guidelines based on SMACNA guidelines in EQ Credit 3.1, Construction Indoor Air Quality Management Plan: During Construction. Note there are other requirements for meeting this credit in addition to SMACNA guidelines.

The mechanical engineer typically provides an indoor air quality management plan that incorporates and/or references the Sheet Metal and Air Conditioning Contractors' National Association (SMACNA) guidelines. The general contractor and subcontractors must comply with these guidelines during construction.

72. ***The answer is*** **(A)** decreasing the lighting power density
(C) decreasing the solar heat gain coefficient

USGBC defines the lighting power density (LPD) as the installed lighting power, per unit area. This measurement is typically given in watts per square foot. A lower LPD will help reduce the energy consumed by the building. The solar heat gain coefficient (SHGC) measures how well a window blocks heat from the sun from entering the building. Decreasing the SHGC means less heat is enters the building through the windows. In a building constructed in a warm climate, this reduces the cooling load.

73. ***The answer is*** **(D)** no credit

USGBC requires the owner or responsible party to purchase for a two-year period a minimum of 35% of the buildings' power from a Green-e renewable source as defined by the Center for Resource Solutions (CRS) Green-e product certification requirements. It is important to note the difference between renewable energy in EA Credit 3, Renewable Energy, which relates to power created onsite, and EA Credit 6, Green Power, which relates to power purchased from a supplier that meets the Green-e definition.

74. ***The answer is*** **(A)** calculated hypothetical baseline energy performance of the systems under measured post-construction operating conditions

USGBC offers two options for complying with EA Credit 5, Measurement and Verification. Both are based on the IPMVP referenced standard, which describes how to develop a theoretical energy-use baseline. The IPMVP Option B uses measurement and verification at the system, or Energy Conservation Measure (ECM), level (rather than the whole-building level). Option B is suited for smaller buildings, like the community center described in this problem.

While the other choices are baseline standards or referenced standards for other credits, they are not correct for this question. Answer choice (B) references Option D of the IPMVP in EA Credit 5, which addresses measurement and verification at the whole-building level. Answer choice (C) references the minimum energy standard for EA Credit 1, and answer choice (D) references the baseline standard for EA Credit 6, Green Power.

75. ***The answer is*** **(B)** fire stopping sealants
(D) aerosol adhesives

EQ Credit 4.1 requires that all adhesives and sealants used on the interior of the building comply with South Coast Air Quality Districts Rule #1168, and that aerosol adhesives comply with GS-36. For EQ Credit 4, LEED defines the interior of the building as the area inside of the weather-proofing system. Roofing and stucco adhesives would be outside of the weather-proofing system.

76. ***The answer is*** **(B)** thirty uses at 1.6 gallons

Based on the data in the *LEED-NC Reference Guide*, a FTE female occupant uses a conventional water closet three times each day. Therefore, 10 females make up 30 uses in one day. A conventional water closet uses 1.6 gallons per flush. The *LEED-NC Reference Guide* provides a table with occupant water use default values and a table on Energy Policy Act of 1992 water use standards. These tables provide the data for determining the water use reduction from baseline standards in WE Credit 3, Water Use Reduction.

77. ***The answer is*** **(C)** The CIR and ruling must be submitted with the LEED application

The USGBC states that the Credit Interpretation Request (CIR) and ruling process was established for project applicants seeking technical and administrative guidance on how LEED credits apply to their projects and vice versa.

A Credit Interpretation Ruling does not guarantee a credit award. The LEED Certification Application must still demonstrate and document achievement of credit requirements in order for credits to be awarded. A credit award is achieved by submitting both the CIR and the ruling when submitting the LEED application. When submitting a CIR, do not format it as a letter. Include only the inquiry and essential background and/or supporting information that provide relevant project details. It is not necessary to include the credit name or contact information since the LEED On-line database automatically tracks this data. Review the document entitled "Guidelines for CIR Customers" on www.usgbc.org for complete CIR submission requirements.

78. ***The answer is*** **(B)** payment in the amount of $250 per credit or prerequisite

Payment should be included with the appeal submission, but the fee is $500 per credit or prerequisite appealed, not $250 as stated in option (B). The project team can pursue an appeal if they feel a credit was incorrectly denied in the Final LEED Review. A team assembled by the USGBC will complete an Appeal LEED Review within 30 days of receipt of the appeal request. Review the information on www.usgbc.org under "Certification Process" for complete appeal guidelines.

SSc 2
4.1
4.2
4.3
4.4
5.1
5.2
7.1
7.2

S.S.5.1

Const

I Green field,
Do not develop (limit of
40 ft from building perimeter
10 ft from parking, patios, walkways & utilities < 12 in
15 ft from major utilities, curbs, roads
25 ft from porous pavements, fields

O
D
C

II On previously developed site.
→ Vegetate 50% of the open space with
Native / adaptive.
(including building roof bridge)
→ 75% of bonus.

+1

→ If the project meets SSc 2, then only
20% of the open space including balcon

5.2

Design

O
D

Maximize open space

If there is a local code
provide open space ≥ 25% greater than local

If no local
= Building footprint

+1

If local code = 0
= 20% of site area

Double

~~S.S.~~

D

We C 1		50%	→	D
We C ~~2~~.1		100%	→	O / D
2	→	50% → 100% (EP)	→	D
3.1	→	20%		D
3.2	→	30% → (40%) (EP)		D

79. ***The answer is*** **(A)** SS Credit 5.2, Site Development: Maximize Open Space
(B) WE Credit 2, Innovative Wastewater Technologies
(D) EQ Credit 8.2, Daylight and Views: Views for 90% of Spaces

The *LEED-NC Reference Guide* states that Innovation in Design points for exemplary performance are not available for EA Credit 1, Optimize Energy Performance nor EA Credit 6, Green Power. Among others, the following credits are eligible for exemplary performance and are listed with requirements.

- SS Credit 5.2, Site Development: Maximize Open Space—Projects with no local zoning requirements must provide open space equal to two times the building footprint.
- WE Credit 2, Innovative Wastewater Technologies—Projects must demonstrate a 100% reduction in potable water use for sewage conveyance.
- EQ Credit 8.2, Daylight and Views: Views for 90% of Spaces—There is no prescribed threshold for determination of exemplary performance. Evaluation is on a project-by-project basis.

80. ***The answer is*** **(A)** $50,000 worth of new FSC certified wood

The requirement for earning MR Credit 7, Certified Wood is that 50% of the total value of the project's permanently installed new wood must be Forest Stewardship Council (FSC) Certified. All wood that is salvaged or reclaimed need not be included in the total value calculation. Any portion of the wood that qualifies for MR Credit 4, Recycled Content cannot also contribute to earning MR Credit 7.